# Seeds and Plants

## Using Nonfiction to Promote Literacy Across the Curriculum

by Doris Roettger

**Fearon Teacher Aids**
Simon & Schuster Supplementary Education Group

# Teacher Reviewers

Rebecca Busch
San Antonio, Texas

Mary Cass
Grosse Pointe, Michigan

Nora Forester
San Antonio, Texas

Delphine Hetes
Detroit, Michigan

Debbie Kellogg
West Des Moines, Iowa

Editorial Director: Virginia L. Murphy

Editor: Virginia Massey Bell

Copyeditor: Susan J. Kling

Illustration: Anita Nelson

Design: Terry McGrath

Cover Design: Lucyna Green

ISBN 0-86653-982-4

Printed in the United States of America

1.9 8 7 6 5 4 3 2

# A Note from the Author

$\mathcal{C}$hildren have a natural curiosity about the world in which they live. They are intensely interested in learning about real things, real places, and real people. They also enjoy and learn from hands-on experiences. Nonfiction books and magazines provide opportunities for children to explore their many interests and extend their base of knowledge.

Reading nonfiction materials is different from reading picture or storybooks. To be effective readers, children need to learn how to locate information or find answers to their many questions. They also need to learn to think about and evaluate the accuracy of any information presented. Finally, they need opportunities to learn the relationship between what they read and the activities in which they apply their new knowledge.

You, as the teacher, can provide opportunities for children to learn from their observations, their reading, and their writing in an integrated language-arts approach across the curriculum.

Modeling thinking strategies and then providing practice across the curriculum will help students become observers and explorers of their world, plus effective users of literacy skills. Encouraging children to extend and demonstrate their understanding through a variety of communication areas—speaking, reading, drama, writing, listening, and art—is also very valuable.

The suggestions in this guide are action-oriented and designed to involve students in the thinking process. The activities do not relate to any one single book. Instead, the strategies and activities are designed to be used with any of the books suggested in the bibliography or with books found in your own media center. The suggested interdisciplinary activities can also be used across grade levels.

Each lesson begins with the reading of a nonfiction book, book chapter, or magazine article—any title that relates to the follow-up activities. During the activity phase and at other class times, students are encouraged to return to the nonfiction selections available in the classroom to find answers to their questions, compare and verify their observations, and add any new information to their current knowledge base.

The individual theme units are designed to be used for any length of time—from a few days to a month or more, depending on the needs and interests of your students.

Suggested goals for this unit are provided near the beginning of this guide on page 16. The webs on pages 6-8 give you an overview of the areas in which activities are provided.

On each page of this guide, there is space for you to write reflective notes as well as ideas that you want to remember for future teaching. This guide is designed to be a resource from which you make decisions and then select the learning experiences that will be most appropriate for your students.

*Doris Roettger*

# Contents

# Literacy Skills

*T*he following literacy skills are addressed in the
*Seeds and Plants* theme guide.

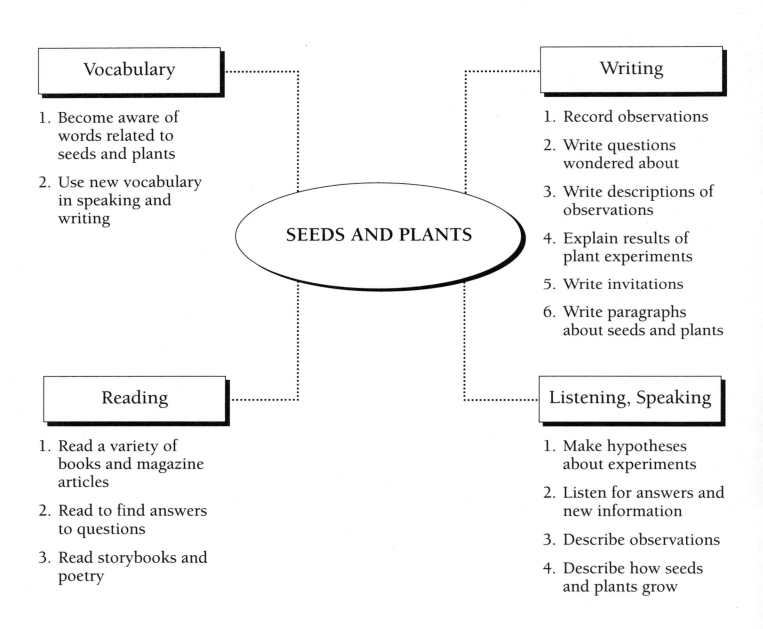

### Vocabulary

1. Become aware of words related to seeds and plants

2. Use new vocabulary in speaking and writing

### Writing

1. Record observations

2. Write questions wondered about

3. Write descriptions of observations

4. Explain results of plant experiments

5. Write invitations

6. Write paragraphs about seeds and plants

**SEEDS AND PLANTS**

### Reading

1. Read a variety of books and magazine articles

2. Read to find answers to questions

3. Read storybooks and poetry

### Listening, Speaking

1. Make hypotheses about experiments

2. Listen for answers and new information

3. Describe observations

4. Describe how seeds and plants grow

# Interdisciplinary Skills

$\mathcal{T}$he following interdisciplinary skills are addressed in the *Seeds and Plants* theme guide.

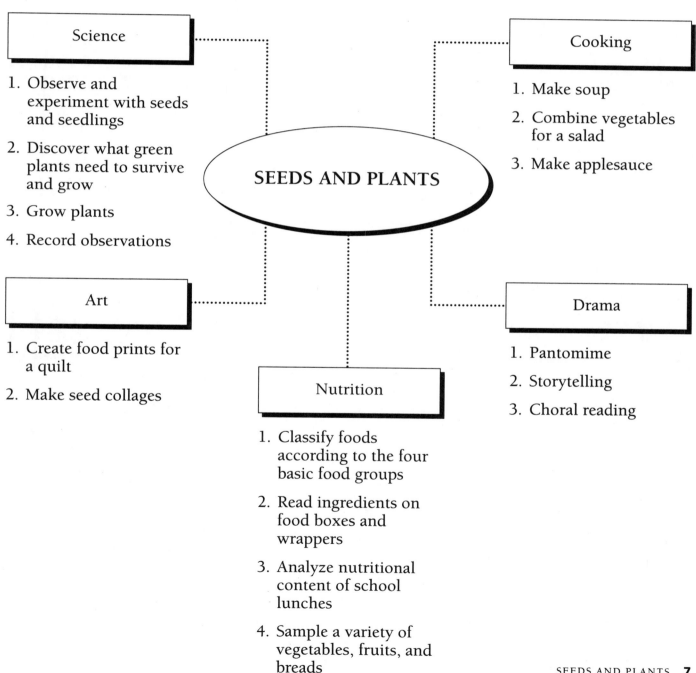

**Science**

1. Observe and experiment with seeds and seedlings

2. Discover what green plants need to survive and grow

3. Grow plants

4. Record observations

**Art**

1. Create food prints for a quilt

2. Make seed collages

**SEEDS AND PLANTS**

**Cooking**

1. Make soup

2. Combine vegetables for a salad

3. Make applesauce

**Drama**

1. Pantomime

2. Storytelling

3. Choral reading

**Nutrition**

1. Classify foods according to the four basic food groups

2. Read ingredients on food boxes and wrappers

3. Analyze nutritional content of school lunches

4. Sample a variety of vegetables, fruits, and breads

# Learning and Working Strategies

$\mathcal{T}$he following learning and working strategies are addressed in the *Seeds and Plants* theme guide.

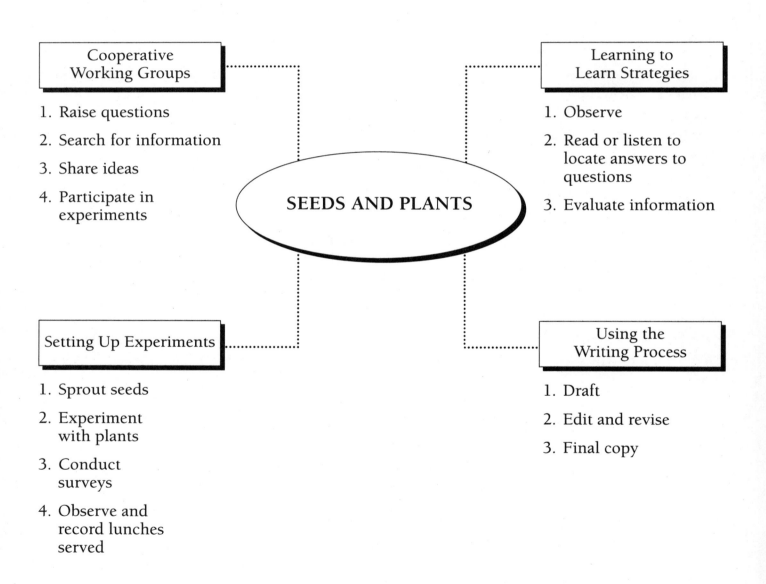

**Cooperative Working Groups**

1. Raise questions
2. Search for information
3. Share ideas
4. Participate in experiments

**SEEDS AND PLANTS**

**Learning to Learn Strategies**

1. Observe
2. Read or listen to locate answers to questions
3. Evaluate information

**Setting Up Experiments**

1. Sprout seeds
2. Experiment with plants
3. Conduct surveys
4. Observe and record lunches served

**Using the Writing Process**

1. Draft
2. Edit and revise
3. Final copy

# About Seeds and Plants

Seeds are an essential component in the creation of a new plant. Seeds contain an embryo, which is comprised of an immature root and a stem. Seeds also have their own supply of stored food and a protective coating.

Seeds vary greatly in size. The largest seed comes from the coconut tree and weighs 50 pounds. The smallest seed comes from the orchid. Eight hundred thousand orchid seeds weigh only one ounce.

There are two main groups of seeds—enclosed seeds and naked seeds. Seeds, such as apple, peach, peas, corn, wheat, and poppies, are enclosed seeds, while pine trees or cone-bearing plants and most shrubs have naked seeds. Seeds are made up of three parts—the embryo, food storage tissue, and a seed coat.

There are approximately 350,000 different kinds of plants. The smallest plants are called *diatoms*, which are microscopic in size, while the largest plants are the sequoia trees found in California.

Plants make their own food from air, water, and sunlight. Most of the plants discussed in this theme guide are flowering plants. Flowering plants produce flowers, fruits, and seeds. Flowering plants have four main parts—roots, stems, leaves, and flowers. Roots usually grow underground and absorb the water and minerals needed for a plant to grow. Roots also help anchor the plant in the soil. Some plants, such as beets, radishes, and carrots, store their food in their roots.

Stems vary greatly from plant to plant. Plants, such as trees, have large stems. Lettuce and cabbage, however, have short stems and large leaves. Some stems grow underground, as does the stem of the potato plant.

Leaves produce food for the plant through a process called *photosynthesis*. The chlorophyll in the leaves absorbs energy from the sun. Water and minerals from the soil are combined with the sun's energy.

Leaves vary in size from less than one inch long and wide to 50 feet long and eight feet wide.

The reproductive part of a flowering plant is called the *flower*. Flowers grow from buds along the stems of a plant. The number of flowers plants produce will vary. Dandelions produce only one flower, while apple trees produce many.

# Suggested Reading Selections

*A* variety of nonfiction and fiction selections for the primary grades is suggested for use with this theme unit. You will probably want to assemble a collection of materials ahead of time. Or, you may wish to have the students help collect several titles from the library as a group activity. The number and type of selections you and the children read will depend on the length of time you devote to this unit, as well as the availability of titles and the level of your students.

## Nonfiction Picture Books

*Apple Tree* by Barrie Watts. Englewood Cliffs, NJ: Silver Burdett, 1987. Describes the annual cycle of an apple tree through photographs.

*Apples: A Bushel of Fun & Facts* by Bernice Kohn. New York: Parents' Magazine Press, 1976. Describes the history, cultural background, and varieties of apples. Recipes and index.

*Bean and Plant* by Christine Back. Englewood Cliffs, NJ: Silver Burdett, 1986. An illustrated book in full color describes how beans grow.

*Bread* by Dorothy Turner. Minneapolis: Carolrhoda Books, Inc., 1989. Describes how bread is produced, its history, and cultural significance. Includes recipes, photographs, glossary, and index.

*Eating* by John Gaskin. New York: Franklin Watts, Inc., 1984. In simple language with color illustrations, this book describes the different types of foods, why we need food, and what happens when we eat. A few projects and glossary included.

*How to Grow a Jelly Glass Farm* by Kathy Mandry. New York: Pantheon Books, 1974. Humorous, colorfully illustrated guide to growing fourteen indoor plants. Simple instructions. Options given to children if the plant doesn't grow.

*Junk Food—What It Is, What It Does* by Judith S. Seixas. New York: Greenwillow Books, 1984. A simply stated introduction to junk food, what is in each food, and how each affects the body.

*A Kid's First Book of Gardening* by Derek Fell. Philadelphia: Running Press, 1990. Presents information on soil, seeds, easy-to-grow flowers, flowers that keep blooming, bulbs, vegetables, fruits, trees, shrubs, houseplants, gardening in containers, and unusual plants.

*Meat* by Elizabeth Clark. Minneapolis: Carolrhoda Books, Inc., 1990. Describes where meat comes from, different types of meats, and how animals are raised that provide the meat we eat. Includes recipes, photographs, glossary, and index.

*Milk* by Dorothy Turner. Minneapolis: Carolrhoda Books, Inc., 1989. Discusses several milk-producing animals. Main focus is on the dairy cow, milk production, and pasteurization. Includes easy, illustrated recipes.

*Plant Experiments* by Vera Webster. Chicago: Childrens Press, 1982. An illustrated book of simple, easy-to-follow experiments with plants.

*Potato* by Barrie Watts. Englewood Cliffs, NJ: Silver Burdett, 1987. Describes the development of the potato from a shoot to a plant. Simple terms, illustrations, and photographs.

*Potatoes* by Dorothy Turner. Minneapolis: Carolrhoda Books, Inc., 1989. Discusses cultivation, history, and nutritional value of potatoes. Includes recipes, activities, photographs, glossary, and index.

*Roots Are Food Finders* by Franklyn M. Branley. New York: T. Y. Crowell Co., 1975. A Let's Read-and-Find-Out Book. Explains in simple terms the function of roots. Includes experiments and illustrations.

*Science Fun with Peanuts and Popcorn* by Rose Wyler. Englewood Cliffs, NJ: Julian Messner, 1986. Experiments with peanuts and popcorn seeds for the classroom and for home.

*Seeds* by Terry Jennings. New York: Gloucester Press, 1988. Colorfully illustrated book about what seeds are, what they need, and what they grow into. Experiments and glossary.

*Seeds: Pop Stick Glide* by Patricia Lauber. Southbridge, MA: Crown, 1981. Detailed introduction to the ways in which plants disperse their seeds. Well-photographed, with simple text.

*Vegetables* by Susan Wake. Minneapolis: Carolrhoda Books, Inc., 1990. Describes different types of vegetables, how they are grown, their history, and how they affect human beings and their health. Includes recipes, photographs, glossary, and index.

*Your First Garden Book* by Marc Brown. Boston: Little, Brown and Company, 1981. A beautifully illustrated book with projects for beginners on how to sprout seeds, grow plants, and care for a garden.

## Poetry and Fiction Picture Books

*Anna's Garden Songs* by Mary Q. Steele. New York: Greenwillow Books, 1989. A book of short poems about summer and the garden. Illustrated.

*Blueberries for Sal* by Robert McCloskey. New York: Puffin, 1976. A wonderful story about a little girl and a bear cub who both wander away from their mothers while picking blueberries.

*Bread and Jam for Frances* by Russell Hoban. New York: Harper & Row, 1964. Frances only wants two kinds of food—bread and jam. But when that's all she ever gets, she becomes unhappy.

*Cloudy with a Chance of Meatballs* by Judith Barrett. New York: Crowell Junior Books, 1976. Food drops from the sky and delights young readers!

*Corn Is Maize: The Gift of the Indians* by Aliki. New York: Crowell Junior Books, 1976. A description of how corn was found by Indian farmers thousands of years ago.

*Eat Your Peas, Louise!* by Pegeen Snow. Chicago: Childrens Press, 1985. A picture book story in rhyme.

*The Gingerbread Boy* retold and illustrated by Paul Galdone. Boston: Houghton Mifflin, 1983. A funny, beautifully illustrated retelling of the adventures of a runaway gingerbread boy.

*Gregory, the Terrible Eater* by Mitchell Sharmat. New York: Macmillan, 1980. An illustrated storybook about a little goat named Gregory who likes to eat fruits and vegetables rather than clothing, paper, and shoes.

*Jack and the Beanstalk* by Paul Galdone. Merlin, OR: Clarion, 1982. Jack climbs the great beanstalk that grows from a bean he bought and confronts a giant at the top.

*Johnny Appleseed: A Tall Tale* by Steven Kellogg. New York: Morrow Junior Books, 1988. A story about John Chapman, known as Johnny Appleseed, who spread cheer and apple seeds in the 1700s. Beautifully illustrated by Steven Kellogg.

*Miss Rumphius* by Barbara Cooney. New York: Viking Penguin, 1982. Alice Rumphius wanted to travel the world when she grew up and then live by the sea. But her grandfather tells her she must do something to make the world more beautiful.

*Mouse Soup* by Arnold Lobel. New York: Harper & Row, 1983. A mouse cleverly obtains ingredients for a delicious soup.

*Pancakes for Breakfast* by Tomie DePaola. San Diego: Harcourt Brace Jovanovich, 1978. A picture book without words about a woman who wants to make pancakes for breakfast.

*The Popcorn Book* by Tomie DePaola. New York: Holiday House, 1978. A wonderfully illustrated picture book of popcorn facts.

*Potatoes, Potatoes* by Anita Lobel. New York: Harper & Row, 1984. A simply told story with a strong message. An old woman on a potato farm has two sons who become enemies in war. She will not feed either of the armies until all the soldiers put down their guns.

*Stone Soup: An Old Tale* by Marcia Brown. New York: Scribner's, 1975. The story of three hungry soldiers who are able to conjure up a feast by making soup from three stones.

*Stone Soup* by Ann McGovern. New York: Scholastic, Inc., 1986. A hungry young man, who is refused a meal by an old woman, makes soup from a stone.

*The Story of Johnny Appleseed* by Aliki. New York: Prentice Hall, 1963. The story of Johnny Appleseed, who spread apple seeds and love throughout the country.

*The Tiny Seed* by Eric Carle. New York: T. Y. Crowell Co., 1970. A picture book telling the story of a seed's life.

*Tony's Bread* by Tomie DePaola. New York: G. P. Putnam's Sons, 1989. A baker loses his daughter, but gains a bakery in Milano, after meeting a determined nobleman and baking a unique loaf of bread.

*The Turnip* by Janina Domanska. New York: Macmillan, 1969. A picture book about a wildly growing turnip.

## Teacher Reference

*Grocery Store Botany* by Joan Elma Rahn. New York: Atheneum, 1974. Explains how plants provide us with food. Includes some experiments and the nutritional value of several different plants.

*Projects with Plants* by Seymour Simon. New York: Franklin Watts, Inc., 1973. Experiments exploring the world of living plants—from houseplants to mold. Provides information on plant processes, such as how water travels through stems and leaves, effects of sunlight versus artificial light, as well as how to grow and care for plants.

## Magazine Articles

"Adventures of Ranger Rick: Saving Endangered Plants" from *Ranger Rick,* National Wildlife Federation, June 1990.

"Bring Spring Indoors" from *Your Big Backyard,* National Wildlife Federation, March 1990.

"Cooking with Cactus Fruits" from *Ranger Rick,* National Wildlife Federation, November 1989.

"How Does Your Garden Grow?" from *Kid City,* Children's Television Workshop, April 1989.

"Let's Grow a Carrot Top Garden" from *Your Big Backyard,* National Wildlife Federation, February 1991.

"Mark Plotkin: Jungle Plant Scientist" from *Ranger Rick,* National Wildlife Federation, February 1990.

"Terrarium" from *Your Big Backyard,* National Wildlife Federation, March 1989.

# Instructional Goals

*I*nstructional goals for this theme unit are provided here. Space is also provided so that you may fill in your own individual goals where appropriate as well. By the end of this theme unit, students should be able to:

1. Name all they know and think they know about seeds and plants.

2. Observe seeds and plants as they grow.

3. Distinguish between fiction and nonfiction selections.

4. Identify the roots, stem, flower, and seeds of plants.

5. Know what plants need to live and grow.

6. Explain and demonstrate how plants grow.

7. Use new vocabulary in their writing and speaking.

8. Classify foods according to the four basic food groups.

9. Raise questions about seeds and plants based on their own curiosity.

10. Find information to answer their questions.

11. Write about their observations.

12. Extend their learning through cooking.

13. Share what they have learned through art activities.

14. Evaluate information they find and decide whether it is the information they wanted to know.

15. ..............................................................................................................

16. ..............................................................................................................

17. ..............................................................................................................

18. ..............................................................................................................

19. ..............................................................................................................

20. ..............................................................................................................

# Getting Started

# Strategies to Raise Curiosity

The following activities are designed to help launch the *Seeds and Plants* theme unit. You may want to use all of the activities or only one or two, depending on the needs of your students. At the beginning of each lesson, reading a nonfiction book or magazine selection to the class serves as a motivator and helps students become more familiar with and involved in using nonfiction selections. For the activities in this section, a general selection on seeds and edible plants would probably be most appropriate. You'll also want to provide plenty of opportunities for children to return to nonfiction selections independently during the activity phases and at other times during class periods as well.

# 1. Classifying Seeds and Non-Seeds

a. Collect a variety of seeds and a variety of items that resemble seeds, such as raisins, M & M's, vermiculite, red hots, and peppercorns. Mix everything together.

b. Divide the class into student pairs. Give each student pair a handful of the mixture. Encourage the children to sort through the mixture separating the seeds from the non-seeds. Have the children place the seeds in small paper cups and save them for later use. Be sure to keep some of the mixture for discussion.

c. Encourage the children to participate in a class discussion answering the following questions: How can you tell which items are seeds? What is a seed? What do you do with seeds?

## 2. Collecting a Variety of Fruits, Vegetables, and Seeds

a. Collect a variety of fruits and vegetables and corresponding seeds. Some suggestions for this activity are listed for you. You might wish to buy packages of the carrot, lettuce, and green bean seeds as well. Some fruits and vegetables not native to your area would be good to include, too.

> carrots and carrot seeds
>
> leaf lettuce and lettuce seeds
>
> lima beans and lima bean seeds
>
> green beans and green bean seeds
>
> apples and apple seeds
>
> oranges and orange seeds
>
> potatoes and cut-up potatoes, each piece containing an eye
>
> peach and a peach pit
>
> avocado and an avocado seed
>
> squash and squash seeds
>
> melon and melon seeds

b. Set up three or four displays around the classroom so that all children have easy access to the different fruits, vegetables, and seeds on display.

c. Have children count the number of seeds in several of the fruits and vegetables. Encourage children to draw conclusions about numbers of seeds in each sample.

d. In a large group, ask children to name and describe the fruits and vegetables on display. Discuss which fruits and vegetables have seeds we can eat. Finally, encourage students to explain how they think fruits and vegetables come from seeds.

## 3. Finding Out What Children Already Know

Create webs of what the children already know about plants and seeds. Begin by writing *What We Know, Questions We Have*, and *What We Have Learned* on large display cards. Post the webs where children can refer to them. As the unit progresses, write the additional information on cards and connect the individual cards to the web with yarn.

# Finding Information in Books and Magazines

For each of the following activities, first select a nonfiction book or magazine and then demonstrate how to find information by thinking aloud and having children work through the process with you. Share what you are doing so that children can learn the thought processes as they learn the strategies. Gradually, ask students to think aloud as they locate answers to their own questions as they read as well.

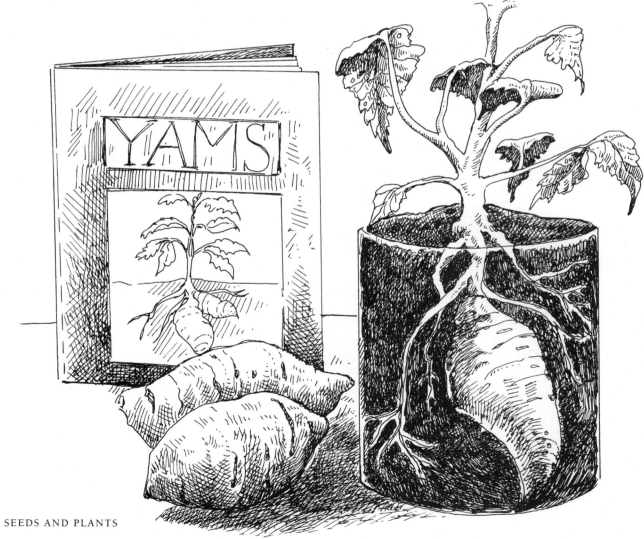

# 1. Locating Information

a. Help children find books and magazines that pertain to seeds and plants.

b. For more advanced readers, model ways children can use a table of contents. Think aloud so students know what you are doing. Then give groups practice identifying titles, page numbers, and locating the pages in books and magazines. Have children think aloud so you know that they understand the process. Repeat the process using an index.

# 2. Finding Answers to Questions

a. Demonstrate how students can find answers to questions by modeling how you would locate an answer to a question posed by one of the children. Again, think aloud as you find the answer. Give students practice in finding an answer to a question. Ask children to think aloud as they find their answers.

b. Encourage students to read books of their choice or have older students read several books to a cooperative working group. You might also make tapes of appropriate selections to place in a listening center. Encourage children to watch or listen for new and interesting facts, as well as for answers to their questions.

c. Read aloud to the students selections from a number of books suggested in the bibliography. After reading each selection, have children recall the new information they have learned. Record their responses in a web format.

# Real-Life Laboratory

# Setting Up Seed and Plant Centers

*S*etting up and maintaining seed and plant centers will give young children an opportunity to identify the different parts of seeds and plants and develop an awareness of the differences among the various types. Begin the lesson by reading aloud a nonfiction selection on the seeds or plants presented in the activities.

# 1. Identifying the Parts of a Seed

Encourage children to find a reading selection that shows the different parts of a seed. If such a book isn't available or is inappropriate for the grade level, students can use the following steps:

a. Soak several kinds of bean seeds in water overnight.

b. Give two or three beans to each child. Have children very carefully open the seeds and find the seed coat, food, shoot, and root.

c. Have children draw and label their own diagrams in individual seed-and-plant booklets.

# 2. Seed Expansion

This activity gives children an opportunity to observe seeds expanding. If possible, begin this activity early in the morning.

a. Fill a small jar with bean seeds or unpopped popcorn. Place a small plastic dish or lid on top of the jar. Fill a second small jar of the same size with beans or popcorn and then fill the jar with water. Place a small plastic dish or lid on top of this jar, too.

b. After several hours, ask the children to observe any changes that have taken place in the two jars.

c. The next day, ask the children to look at the two jars again. Have students speculate why the bean seeds or popcorn are changing in the jar with water.

# 3. Seed Collection

Begin a class seed collection. Wrap several sets of seeds in cellophane and label each packet. Place the seed packets in an observation center. Ask the children to look for differences and similarities among the packets.

# What Green Plants Need to Survive and Grow

$\mathcal{T}$he following activities will give children the opportunity to perform experiments on several different kinds of green plants. Begin the activities by reading aloud several appropriate nonfiction selections.

# 1. Watching Bean Plants Grow

You will need several containers for this activity. Fill each container with a fifty percent potting soil-vermiculite mixture. Have children plant two or three bean seeds about 1/4 inch deep in each carton. Water the soil and keep moist. Have children record what they plant in each container on a large classroom calendar.

a. After the seeds have sprouted, encourage children to pose some "I Wonder" questions about what plants need to survive and grow. For example,

"I wonder if plants grow better in the sun or in the shade?"

"I wonder what happens if we water some plants and not others?"

b. Record the children's questions and observations on a large piece of butcher paper in a web format. Continue to add to the web as children learn additional information and make new observations.

# 2. Experimenting with Green Plants

This activity helps children realize that green plants need sunlight, air, and water to survive and grow. These experiments are especially good for cooperative working groups. Children can compare their results with those of the other groups.

a. You will need two plants of the same size and type. Put one plant in sunlight and the other in a closet. Water both. Have children observe the plants and once a week record how each plant looks.

b. Place two plants of the same size and type in a sunny location in the classroom, but put one plant in a plastic reclosable bag. Water both. Seal the plastic bag, except for a small opening. Insert a straw into the opening and draw as much air out of the plastic bag as possible. Quickly seal the bag closed. Have children observe the plants and once a week record how each plant looks.

c. Put two plants of the same size and type in an area where they both get sunlight. Water one plant and not the other. Once a week, have children observe and record the condition of the two plants.

d. After the experiments are over, have children draw some conclusions about what plants need to survive and grow. Record the children's observations on webs displayed in the classroom.

## 3. Observing Root Growth

This activity helps children study roots and find out how plants get water and other nutrients. Before beginning, read aloud an appropriate nonfiction selection.

a. Divide the class into cooperative working groups. Provide each group with one or two clear plastic cups filled with a fifty percent potting soil-vermiculite mixture. Have students plant three bean seeds close to the side of each cup.

b. As the roots grow, have children use a magnifying glass to observe the root hairs. Ask students to describe what they see. Point out to the children that plants get their nutrients and water through roots and root hairs.

## 4. Experimenting with Celery

Children will be using celery stalks to study stems in this activity. You might wish to have several different cooperative working groups perform this experiment. Begin the activity by reading a nonfiction selection about plant stems.

a. Cut at a slant across a leafy stalk of celery near the bottom. Make a lengthwise cut from the bottom to within an inch of where the leaves branch out. Fill two glasses half full of water. Put 1 teaspoon of blue food coloring in one glass so that the water has a strong tint. Add 1 teaspoon of red food coloring in the second glass. Put each section of the celery stalk into a glass. Place the glasses and the celery in a sunny spot in the classroom.

b. Look at the celery stalk after a few hours. Does the stalk look different? Leave the stalk in the glasses overnight. In the morning, take the stalk out of the glasses and cut across each section. What colors are visible in the sections? Try to pull out a colored string to see where it goes.

c. Point out to the children that the celery stalk is actually a stem. The strings are the tubes that carry the nutrients and the water the plant needs from the roots to the leaves. Then point out the veins in the leaves. The veins in the leaves connect with the tubes in the stems.

## 5. Conducting Independent Experiments

Have children look for additional plant experiments in the nonfiction selections you have collected for the classroom. Provide help as needed for conducting new experiments. Compare experiments done in the class with similar experiments found in books and magazines.

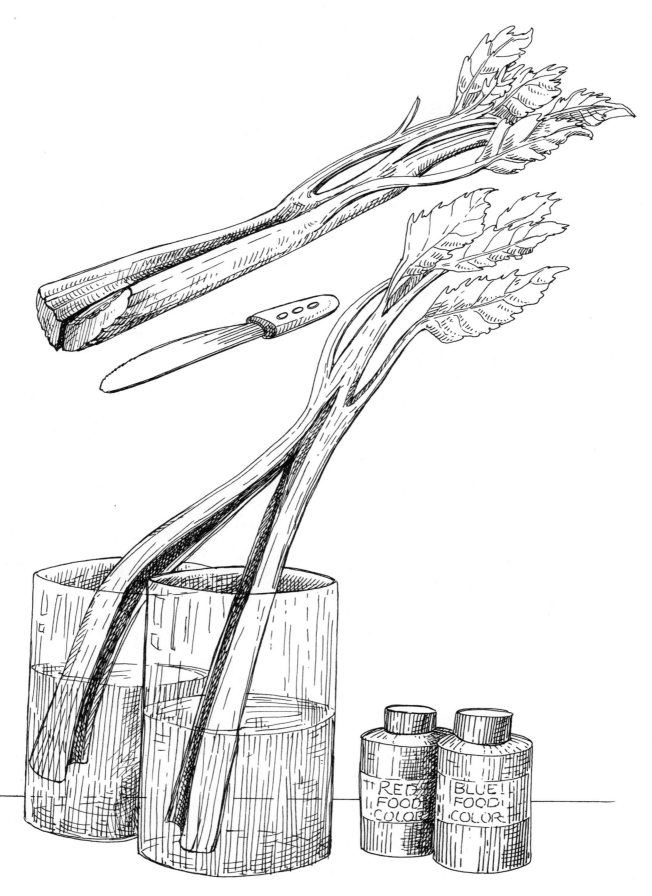

| Sunday | Monday | Tuesday | Wednesday | Thursday | Friday | Saturday |
|--------|--------|---------|-----------|----------|--------|----------|
|  | 1<br>Seeds<br>Planted | 2<br>No<br>Change | 3<br>Small<br>root<br>Showing | 4 | 5 | 6 |
| 7 | 8 | 9 | 10 | 11 | 12 | 13 |
| 14 | 15 | 16 | 17 | 18 | 19 | 20 |
| 21 | 22 | 23 | 24 | 25 | 26 | 27 |
| 28 | 29 | 30 |  |  |  |  |

# 6. Sprouting Seeds

Begin the lesson by reading aloud a nonfiction selection on sprouts. Or, you might read a selection on bean, pea, or popcorn plants.

Materials needed:

> wide-mouth jars or reclosable plastic bags
>
> paper towels
>
> cotton balls
>
> lima beans, peas, and popcorn seeds
>
> classroom calendar with large spaces for drawing different plant stages

Procedure:

1. Line the jars or bags with several thicknesses of paper towels.

2. Place cotton balls inside the jars to hold the paper towels in place (cotton is not needed in the plastic bags).

3. Pour enough water in the bottom of the jars or bags so the paper towels soak the water up and remain damp.

4. Place seeds between the glass or plastic bag and the paper towels so the seeds are visible.

   a. Divide the class into cooperative working groups. Have the students place two bean, two pea, and two popcorn seeds in their sprouting jars or bags. Place all jars or plastic bags in a window for light.

   b. Have children observe and record the growth of the sprouts on a classroom calendar and in individual seed-and-plant booklets. Discuss and determine the following points:

      ✦ whether all the seeds begin sprouting on the same day

      ✦ how many days it takes from the first seed to sprout to the last

      ✦ how many seeds do not sprout

      ✦ whether all the sprouts look the same

## 7. Measuring Root Growth

This activity helps children become aware of how fast roots actually grow. You'll want to begin the activity by reading aloud a short selection about roots. Refer to the materials and the procedures used in Activity 6, if appropriate.

a. Have the children in each cooperative group select a seedling that has just sprouted from the previous activity and measure the length of the root with a ruler. Then have children draw a picture of the root and record its length in individual seed-and-plant booklets.

b. After three days, have students measure a root from the previous activity again. Ask the cooperative working groups to look at the lengths of all the roots and tell whether all have grown the same amount in three days. What are the differences between the sprouts in the various jars? What are the differences between the beans, peas, and popcorn?

c. Have children record the length of the roots on a classroom calendar and in individual seed-and-plant booklets.

## 8. Direction of Root Growth

a. To find out which way roots grow, put several wet paper towels on Styrofoam trays.

b. Place four bean or popcorn seeds on the wet paper towels. Put plastic wrap or clear plastic bags over the seeds so the water doesn't leak out. Stand the tray on one end.

c. Two days after the seeds have sprouted, turn the tray on its side. Every two days, turn the tray again. Have the children describe what happens to the roots each time the tray is turned.

# 9. Transplanting Seedlings

To find out why seedlings need to be transplanted, each cooperative working group will need some leafy seedlings sprouted in class. Before beginning the activity, read aloud a nonfiction selection about transplanting seedlings, if a selection is available.

a. When the seedlings have lots of leaves, ask the children to select two of the seedlings to transplant.

b. Plant each seedling in a container filled with a fifty percent potting soil-vermiculite mixture. Set the container on a dish and water the soil.

c. Have students continue to observe the seedlings that have not been transplanted. Ask the children to speculate why the seedlings that were not transplanted die, while the ones that are transplanted live. Record the children's observations.

## 10. Raising Bean Sprouts

To raise bean sprouts to eat, you'll need 1/2 cup mung bean seeds, one quart jar with a screw-top lid, and water. Encourage students to find a reading selection that provides directions for growing sprouts to eat. If such a selection isn't available, children can use the following steps:

a. Wash beans and soak them in water overnight.

b. Drain off the water. Punch holes in the jar lid top using nails and a hammer.

c. Put the beans in the jar, screw on the lid, and place the jar in a dark place. Rinse beans two or three times a day.

d. Beans will sprout in five days and be ready to eat. Serve alone or on bread with peanut butter.

# Cross-Curriculum Activities

# Writing Arena

*W*riting helps children think about what they have observed and what they already know. It also helps them synthesize their thoughts and then communicate their ideas to others. Read a nonfiction selection to the class before beginning a writing activity. You might point out some sentences or paragraphs in the selection that are especially well-written.

## 1. Recording Observations

Record the children's observations about several different plants, how they look, and how they grow on a "Things We Are Observing" web. Later, have children dictate stories from the webs created.

## 2. Recording and Answering Questions

As questions about plants and seeds arise, record them under the heading "Questions We Have and Things We Wonder About" on a "What Do We Already Know?" chart. Help the children write "Question and Answer" booklets from the information gathered.

## 3. Writing Letters

Invite students to write letters to seed companies requesting copies of their seed catalogs. Introduce or review proper letter-writing form. Listed below are a few addresses.

Earl May Seed & Nursery
208 N. Elm St.
Shenandoah, IA 51603

Gurkey's Seed & Nursery Co.
Yankton, SD 57079

Henry Field Seed & Nursery
Shenandoah, IA 51602

Stokes Seeds
1381 Stokes Bldg.
Buffalo, NY 14240

Van Bourgondien Bros.
Box A, Dept. 4330
Babylon, NY 11702

W. Atlee Burpee & Co.
Warminster, PA 18974

## 4. Writing a Recipe Booklet

Help children write a recipe booklet, compiling favorite recipes from home. Have children illustrate each recipe as well. Duplicate a booklet for each child. Children can color the drawings and take their booklets home as gifts.

## 5. Tasting Party and Invitations

a. Hold a class meeting to discuss the different foods that come from the plants the children have been growing in school. Choose several foods to be shared at a tasting party.

b. As a group, discuss the information that should be included on a party invitation. For example,

date and time of party

place of party

foods to be tasted

c. Children can write individual invitations to children in other classrooms or write one invitation from the entire class to another class in school. Principals, secretaries, and cafeteria helpers might be invited, too.

## 6. Writing a Big Book and Accompanying Little Books

a. Ask children to think aloud about the ideas they would like to include in a book they would write about seeds and plants. List the ideas as the children name them.

b. Have children reread their ideas and think aloud about how the ideas go together and which ideas they want to write about.

c. As a class, have the children dictate what they want to say in a class big book. Leave space for revisions. Put the dictation aside for a few days.

d. Have the children reread their descriptions. Are the ideas in the right order? Is there anything else they would like to say? Could they say something in a better way? Make all revisions on the original draft.

e. When making a final copy of the big book, leave space on each page for illustration. Invite children to illustrate the book themselves.

f. Have children copy the revised big book story in smaller books, following exactly the same format of the bigger book. Encourage students to copy the illustrations, too.

# Cooking Makes It Memorable

$\mathcal{C}$ooking activities add the sense of taste and smell to the learning experience and help children form lasting impressions of what they have learned. Begin the lesson by reading a book or magazine article about foods aloud to the children. Follow the reading with a discussion of the key concepts. Then select the activities that seem appropriate for your class.

# 1. Making Soup

Here are some books to read that are all about soup.

> *Chicken Soup with Rice: A Book of Months*
> by Maurice Sendak
>
> *Mouse Soup* by Arnold Lobel
>
> *Stone Soup* by Marcia Brown
>
> *Stone Soup* by Ann McGovern

a. Critical-thinking questions about the books are provided for discussion after reading.

1. How is the story in the book *Stone Soup* by Ann McGovern similar to the story by Marcia Brown? How are they different?

2. How did the young man in *Stone Soup* by Ann McGovern get the woman to make soup for him? Do you think he had this planned when he stopped at the woman's house? Why or why not?

b. Conduct a survey among the children. Ask about their favorite soups. Chart the information. Conduct a survey of other classes, too. Chart the information from each group. You may wish to use a different color marker for each group so that the children can see similarities and differences.

c. Have students prepare soup, using a recipe from a favorite cookbook. Or use the recipe provided here.

Ingredients:

> potatoes
> canned tomatoes or tomatoes grown in class
> stalks of celery with leaves
> green beans
> frozen mixed vegetables (optional)
> parsley
> onions (peeled)
> small noodles
> pepper and salt
> cans of beef broth (1 can for every 4 servings)

Directions:

1. Pour cans of beef broth plus cans of water into a large pot.

2. Wash all fresh vegetables, except the onions, under cold, running water.

3. Peel the potatoes and cut into 1/2-inch cubes. Cut the celery and the celery leaves, onion, and all other fresh vegetables into small pieces. If using frozen vegetables, cook the fresh vegetables 15-20 minutes before adding the frozen ones.

4. Cut tomatoes into bite-sized pieces and add to the broth. Add any juice, too.

5. Add 1/2 teaspoon of pepper and a little salt. Simmer the soup for one hour, stirring occasionally. Add the small noodles and cook for 15 minutes.

## 2. Making Applesauce

Several storybooks and nonfiction books relating to apples are suggested for sharing with the class before making applesauce.

*Apple Tree* by Barrie Watts

*Apples: A Bushel of Fun & Facts* by Bernice Kohn

*Johnny Appleseed: A Tall Tale* by Steven Kellogg

*Rain Makes Applesauce* by Julian Scheer

*The Story of Johnny Appleseed* by Aliki

a. Have children compare and contrast the books listed above, particularly the different versions of Johnny Appleseed. How are the stories the same? How do they differ? What parts of the stories could be true? What parts are probably make-believe?

b. Bring in a variety of apples, such as Jonathan, Red and Yellow Delicious, Macintosh, and Granny Smith. Cut the apples into eighths. Have the children sample each one. So that they can tell the difference in flavor, have children take a sip of water between each bite. Conduct a survey of the children's preferences. Chart the results.

c. Before making applesauce, encourage students to find two or three recipes in cookbooks to compare, and then select one. If no recipe is available, the children can follow the directions provided here.

1. Wash one apple for each person. Peel and core the apples and cut into eighths.

2. Cover the apples in a pan with a small amount of water. Add about 1/4 cup sugar for every four medium apples. Cook until apples are tender.

3. Mash the apples, then add 1/2 stick cinnamon for every four apples (or ground cinnamon) to taste.

4. Cool until ready to serve. Remove cinnamon sticks before serving.

# 3. Bread-and-Jam Day

a. For a bread-and-jam day, begin the lesson by reading the following fiction books.

> *Bread and Jam for Frances* by Russell Hoban
>
> *The Gingerbread Boy* by Paul Galdone
>
> *Pancakes for Breakfast* by Tomie DePaola
>
> *Tony's Bread* by Tomie DePaola

b. After reading the books suggested, hold a class meeting to discuss the following questions, plus others.

   1. How many food groups were represented when Frances ate only bread and jam?

   2. What do you think Frances needs to eat to have a balanced diet?

   3. How was Tony's bread different from the breads we eat?

c. Invite children to sample a variety of breads.

   1. Cut up a variety of breads into bite-sized pieces. Suggestions include white, wheat, rye, pumpernickel, English muffins, tortilla, pita, cornbread, bagels, and any other breads brought in by the children.

   2. Have the children sample each type of bread. Ask them to notice the differences among the various breads. Compile a list of characteristics as the children name them.

   3. Conduct a survey of favorite breads and graph the results.

## 4. Vegetable Day

For a special vegetable day, begin by reading the following books.

> *Eat Your Peas, Louise!* by Pegeen Snow
>
> *Potatoes, Potatoes* by Anita Lobel
>
> *The Turnip* by Janina Domanska

a. After reading the fiction books listed, or your own favorites related to vegetables, ask the following critical-thinking questions.

   1. What excuses do you use to not eat your vegetables?

   2. Why are vegetables good for you?

   3. How do turnips grow? Are they more like carrots or beans?

   4. How believable are the stories we just read?

b. Have the children sample all the vegetables that they have grown in their classroom garden. Combine the vegetables into a salad.

## 5. Popcorn Day

A good book to read for a popcorn day is *The Popcorn Book* by Tomie DePaola. Then have fun making and eating pizza-flavored popcorn.

---

**PIZZA-FLAVORED POPCORN**

1/4 cup butter (1/2 stick)

2 quarts unsalted popped corn

1 teaspoon dry Italian salad dressing mix

Melt butter and pour over warm popped corn. Sprinkle with Italian salad dressing mix and serve.

---

# Art All Around

Art activities help children visualize the concepts they have been reading about and discussing. Frequently, through art, children will want to review something they have read, heard, or observed so that their artwork is as accurate as they can make it. You'll want to have several resources available for children to refer to. Select a book or magazine article with good photographs or illustrations to read aloud to the class before beginning any activity or activities.

# 1. Food Prints for a Quilt

To make food prints for a quilt, you will need the following materials:

> potato slices
>
> green pepper rings
>
> orange slices
>
> carrot sticks
>
> paints and paintbrushes
>
> 6" x 6" squares of construction paper

a. Ask children to brush paint on the vegetables and fruits they select and then print a design onto the squares of paper.

b. Create a quilt by having the children arrange the printed squares onto a 3' x 5' sheet of paper in a pleasing pattern. Glue the squares in place. Hang the quilt in the classroom.

# 2. Seed Collages

Create seed collages for refrigerator magnets. Provide the following materials:

> small cardboard or wooden heart shapes
>
> variety of small seeds
>
> glue
>
> small magnets
>
> shellac
>
> paintbrushes

a. Have children create a design with seeds on the cardboard or wooden shapes. When they are satisfied with their designs, have the children glue the seeds in place.

b. After the glue has dried, paint a thin coat of shellac on the front of each shape, let dry, and then paint a thin coat of shellac on the back.

c. Glue a magnet onto the back of each shape and then give away as a gift.

# Vocabulary Fun

The activities in this section are designed to give children practice using the vocabulary they are learning as they observe and study the different seeds and plants in this theme unit.

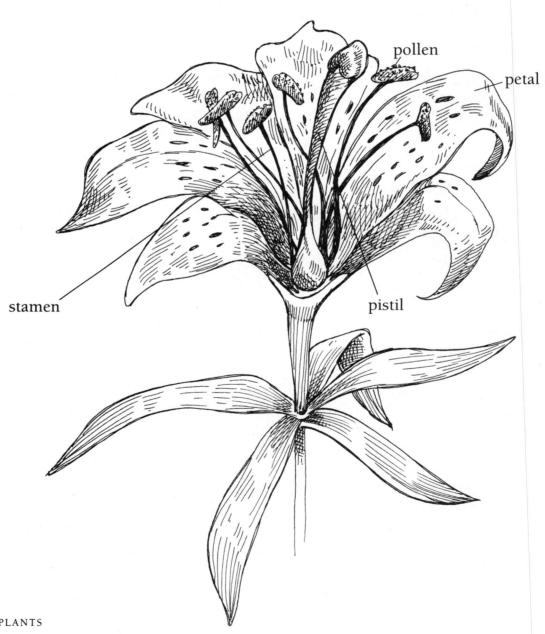

pollen

petal

stamen

pistil

# 1. New Vocabulary

a. List the vocabulary words on chart paper and then display the chart prominently in the classroom. Display pictures of a variety of seeds and plants, too.

b. Have children make their own vocabulary flashcards. Encourage the children to write the vocabulary words on one side of the flashcards and the definitions or descriptions on the other. The flashcards can be used in the students' writing.

**Roots**—grow downward, holding the plant in place in the ground.

**Root hairs**—grow on the root, soaking up water and minerals, both of which are needed by the plant to survive and grow.

**Stem**—connects the roots to the leaves. Water and minerals that are needed by the plant are carried through the stem to the leaves. The food that is created by the leaves is carried through the stem to the roots.

**Leaves**—make food for the plant.

**Flowers**—produce the seeds. Each plant has its own flower. Flowers have petals, a pistil, and stamen.

**Pistil**—is the center part of the flower.

**Stamen**—are located around the pistil. The pollen is located at the top of the stamen.

**Pollen**—is the yellow dust inside the flower. The pollen must reach tiny egg cells inside the flowers to make the seeds. Pollen sticks to the bodies of insects and rubs off when the insects fly from flower to flower in search of food.

**Fruit**—the part of the plant that contains the seeds.

## 2. Concentration Card Game

Make a concentration card game. Write each vocabulary word on a card. Make a definition card for each of the vocabulary words, too. Children match each word with its meaning.

# 3. Big Weed Card Game

Make four sets of vocabulary cards. Or, make two sets of words and two sets of definitions. A pair would then consist of a word with its definition. Also, make one "Big Weed" card.

a. Two, three, or four children may play the game. Use either the set of vocabulary word cards or the word and definition cards.

b. Deal all the cards. Children are to match the pairs—either the vocabulary cards, or the vocabulary cards and definitions. Have players decide what will make a pair before the game begins.

c. The dealer deals all of the cards to the players. The person to the left of the dealer begins play. This person draws from the person to his or her left and then lays down all pairs in his or her hand face down. Play continues with the next person drawing a card from the player on the left and laying down all pairs.

d. Play continues until a player has no more pairs in his or her hand. The person who has the Big Weed card is the loser, and the person with the most number of pairs is the winner of the hand.

# Growing a Classroom Garden

*P*lanting a classroom garden will help children gain a sense of understanding of how some of the foods they eat come from plants they can actually grow themselves. They will have hands-on experiences taking care of the garden from planting to harvesting. Since the plants grow and develop at different rates, planting times should be staggered accordingly. Prepare and eat the vegetables as they are ready. Have special lunches of soup and salad. Begin the activities by reading aloud a nonfiction selection, perhaps one on setting up a vegetable garden. Or, you might read about some of the specific seeds and plants that will be planted. Many of the plantings call for rocks in the bottoms of the containers. Use river rock, not gravel. Gravel is treated with chemicals and may kill the plants as they begin to sprout and grow.

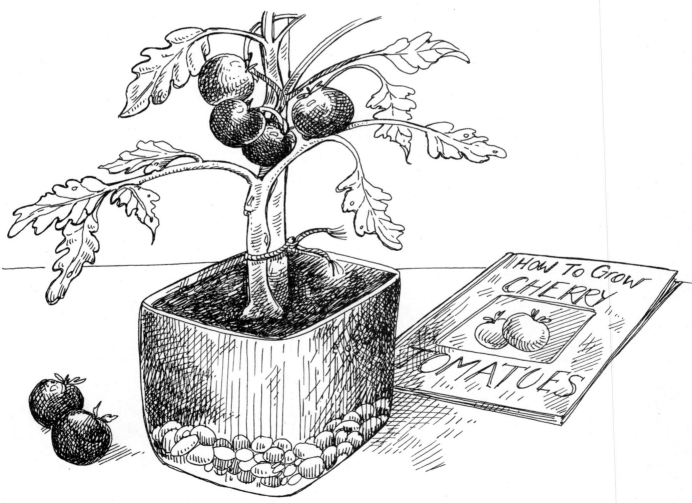

# 1. Growing Potatoes

Growing potatoes in a pail takes approximately 90 days.

a. Cut up firm potatoes into sections with one eye or growth bud in each piece.

b. Put a layer of small rocks on the bottom of a large pail or bucket for drainage. Have children fill the container half full with a fifty percent potting soil-vermiculite mixture. Place the pieces of potato in the soil and cover lightly. Keep the soil moist.

c. Take Polaroid pictures of the plant growth every two weeks. Post the pictures on the bulletin board.

d. When the plants begin to flower, have children study the flowers carefully with a magnifying glass so they can clearly see the pistil, stamen, and the pollen.

# 2. Growing Cherry Tomatoes

Growing cherry tomatoes from seed to a tomato that is ready to eat takes approximately 75-80 days.

a. Use large containers, such as gallon milk jugs with the tops cut off. Have children put a few rocks in the bottom of the container to improve drainage, then fill half full with a fifty percent potting soil-vermiculite mixture. Plant the tomato seeds in the soil and water thoroughly. Keep the soil moist.

b. Take Polaroid pictures of the plant growth every two weeks. Post the pictures on a bulletin board.

c. Have children look carefully at the flowers when they appear. Discuss how the flowers on the tomato plants look different from the flowers on the potato plants. Have children use a magnifying glass to identify the pistil, stamen, and the pollen.

## 3. Growing Carrots

Growing carrots from a seed takes approximately 72 days.

a. Plant carrot seeds 1/4 inch deep in a pail or clay pot. Choose a brand of seed for shorter carrots rather than regular carrots, so there will be plenty of room for growth.

b. Every two weeks, have the children pull out one carrot plant. What changes do they see? How much longer is each carrot that they pull? How much thicker? Have students record the changes.

c. Measure the growth of the stems and leaves each week and record this information on a classroom calendar.

# 4. Planting Beans, Radishes, and Lettuce

Have children plant beans, radishes, and lettuce. Note the growing times listed below and plant the seeds so that the vegetables all mature about the same time. The approximate length of time from planting to maturity for each is:

| | |
|---|---|
| bush green beans | 48-57 days |
| lettuce | 45-55 days |
| radishes | 25-27 days |

a. Have students record when they plant each type of seed. Then have students observe and record when the different seeds first sprout. Measure how tall each plant grows in a week.

b. Take Polaroid pictures of the plant growth each week. Post the pictures on the bulletin board.

c. Ask students to notice which of the plants have flowers. Have children use a magnifying glass so they can clearly see the pistil, stamen, and the pollen. Encourage students to draw pictures of the flowers in individual seed-and-plant booklets.

# 5. Growing a Yam Plant

Have each cooperative group grow a yam plant. Use the plants as centerpieces.

a. Stick three or four toothpicks around the middle of a yam. Suspend the yam by the toothpicks in a wide-mouth jar with the thicker end of the yam extending up out the top. Pour enough water into the jar so that the yam touches the water. Make sure the water is always touching the yam.

b. Have children count the number of days the yam takes to sprout. Record the information on a large classroom calendar.

# The Drama Scene

Drama is a positive, fun, and fulfilling way of learning in which children can practice predicting, planning, organizing, and problem solving. Before beginning the following drama activities, read aloud an appropriate nonfiction selection.

# 1. Pantomime

Have children stretch out their arms, then turn and stretch out their arms again. Ask children to pantomime all their movements within the space around them.

a. Ask children to be a seed by crouching down and taking the appropriate position.

b. Have children show with their arms and legs how a seed starts to grow. As the root grows longer and longer, they should slowly begin to stand up. As the leafy stem gets longer and longer, they should slowly move their arms up into the air. As the seed grows into a plant, the children stand up tall with their arms extended above their heads.

# 2. Storytelling

a. Divide the class into groups of three. Invite children in each group to take turns imagining they are a seed growing into a plant. Encourage students to tell what their roots, stems, and leaves would look like. Have students tell how they eat, what they eat, whether they have any flowers, and what kind of food they will eventually produce.

b. Encourage children to add actions as they tell their stories.

# 3. Poems for Listening and Chanting

a. "Potato" and "Radish" are two fun poems for listening and chanting that can be found in *Anna's Garden Songs* by Mary Q. Steele. Provide several opportunities for the children to read the poems aloud. Invite students to make up actions as they read along.

b. Write a class poem describing the growth of a seed into a plant. Practice chanting the poem together. Encourage children to write poems of their own to share with the class.

# Nutrition

*T*he activities in this section will help children realize how certain plants and foods are necessary for their health and physical growth.

# 1. Reading and Discussing Stories

a. Read the story *Gregory, the Terrible Eater* by Mitchell Sharmat aloud to the class. Ask the children to decide why Gregory was called "a terrible eater." What kinds of foods did Gregory like? Who eats the kinds of foods that Gregory likes?

b. Read *Eating* by John Gaskin, or other books you have that describe the four basic food groups. Have the children name the four basic food groups. List the food groups on a large chart. Leave space on the chart for children to write in the names of various foods that fit into each category. Display the chart in the classroom.

# 2. Food Classification

Set up a food classification center. Send a note home asking children to bring to school empty food containers, such as cereal boxes, egg cartons, milk cartons, macaroni and cheese boxes. You might want to exclude condiments from this activity.

a. Divide the center into four sections, one for each of the four basic food groups. Put together a set of food containers for each cooperative working group. Try to make the sets fairly equal in number and types of containers. Distribute the sets to each group of students.

b. Ask the children to sort the materials in the center according to the appropriate food groups. Any item in question should be discussed and a decision concerning category made based on the listed ingredients.

# 3. Identifying Junk Foods

Read aloud sections from *Eating* by John Gaskin or *Junk Food—What It Is, What It Does* by Judith S. Seixas. Hold a class discussion about what junk foods are and why they are not good for our bodies. Make a chart listing the names of the junk foods the class eats and display the chart in the classroom.

# 4. Graphing Cereal Grains

Read *Corn Is Maize* by Aliki and other books dealing with grains to the children. Ask children to bring empty cereal boxes from home.

a. Distribute the cereal boxes among the cooperative working groups. Ask each group to read the ingredients on the cereal boxes to determine the type or types of grain each cereal contains.

b. Develop a chart listing the name of each cereal under the appropriate heading.

| WHEAT | OATS | CORN | RICE |
|---|---|---|---|
| Crackling Krispies | Cheerios | Corn Flakes Corn Chex | Rice Chex |

c. After all the cereals have been classified, ask children to decide which grain is used the most in cereals. Then graph the children's favorites.

# 5. Analyzing Foods

Analyze the foods served in the school lunchroom for one week. Divide a bulletin board into four sections, one for each of the four basic food groups. Label each section.

a. Write on a card the name of each food served and post it in the appropriate section. Use a different color of marker for each day.

b. At the end of the week, review the foods served.

c. Hold a class discussion about what the children ate during the week. Which type of foods did they like the most? Which did they like the least?

d. Invite a school cook to come speak to the class about how school menus are planned.